F.S.MOOSE

Space Facts Galore

Fun Facts About Space From Our Solar System to the Far Reaches of the Universe

First edition

This book was professionally typeset on Reedsy.
Find out more at reedsy.com

Contents

1

Introduction

Welcome to "Space Facts Galore," a journey that transcends the boundaries of our earthly existence and ventures into the vast, enigmatic realm of space. This book is not just a collection of facts; it is an invitation to explore and understand the wonders that lie beyond our planet.

Purpose of the Book

At the heart of this book is a desire to ignite curiosity and wonder about the cosmos. It aims to present space exploration in a way that is both engaging and informative, making the vast complexities of the universe accessible to readers of all ages. By unraveling the mysteries of space, we hope to inspire a new generation of astronomers, scientists, and dreamers who look up at the night sky with awe and ambition.

Importance of Space Exploration

Space exploration is more than a scientific endeavor; it is a testament to human curiosity and our relentless pursuit of knowledge. It challenges our understanding of life, existence, and our place in the universe. This book highlights the significance of space exploration in advancing technology, enhancing our understanding of environmental changes, and fostering international cooperation.

Overview of What the Universe Holds

Prepare to embark on a cosmic journey that takes you from the familiar craters of our Moon to the farthest galaxies that light up our universe. We'll explore the dynamic solar system, the mysterious black holes, the twinkling stars, and much more. Each chapter unravels the secrets of the cosmos, offering a glimpse into the endless wonders and possibilities that space holds.

Join us as we traverse the universe, one fascinating fact at a time, in "Space Facts Galore."

2

Our Solar System

Overview of the Solar System

Our solar system is a cosmic neighborhood teeming with diverse celestial bodies, each playing a unique role in the vast tapestry of space. Located in the Milky Way galaxy, it comprises the Sun, eight planets, their moons, and a myriad of smaller objects like dwarf planets, asteroids, comets, and meteors. This chapter offers a journey through our solar system, shedding light on each component's distinct characteristics and its place in the cosmic order.

The Sun: Our Star

At the center of our solar system lies the Sun, a colossal star that provides the necessary warmth and light for life on Earth. This massive ball of hot plasma is the ultimate energy source, driving weather, climate, and ocean currents on Earth. The Sun, like other stars, is a fusion reactor where hydrogen atoms fuse to form helium, releasing enormous amounts of energy in the process. Its magnetic field and solar winds have profound effects on the entire solar system.

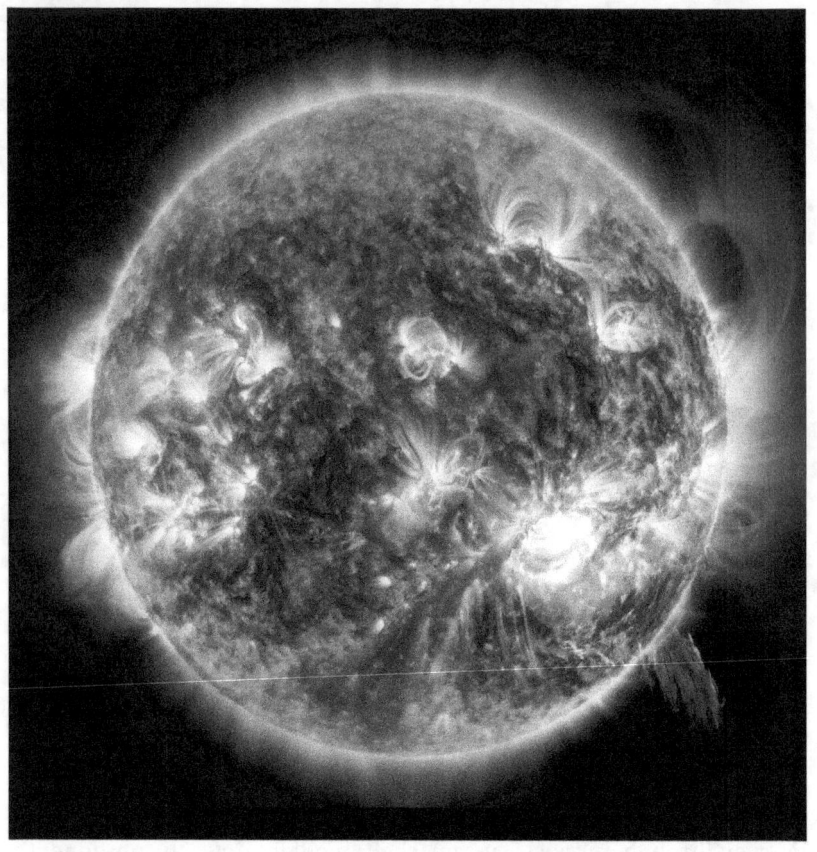

Planets: From Mercury to Neptune

Our solar system is home to eight remarkable planets, each with its own story. Mercury, the closest to the Sun, is a barren, cratered world with extreme temperatures. Venus, shrouded in thick clouds, has a runaway greenhouse effect making it the hottest planet. Earth, our home, is a blue gem with life-sustaining conditions. Mars, the red planet, intrigues scientists with its potential for past life. The gas giants – Jupiter and Saturn – are known for their size, rings, and moons. Uranus and Neptune, the ice giants, are mysterious worlds far from the

Sun, with intriguing atmospheres and moons.

Dwarf Planets and Asteroids

Beyond the main planets, our solar system contains dwarf planets like Pluto, known for its heart-shaped glacier, and Ceres, located in the asteroid belt between Mars and Jupiter. Asteroids, the rocky remnants from the solar system's formation, orbit the Sun mainly in the asteroid belt. These smaller bodies hold clues to the solar system's history and evolution.

Comets and Meteors

Comets, often dubbed as 'dirty snowballs,' are icy bodies that release gas and dust, forming spectacular tails as they approach the Sun. Meteors, commonly known as shooting stars, are bits of space debris burning up upon entering Earth's atmosphere. These ephemeral streaks of light are a favorite among stargazers.

The Mystery of the Oort Cloud

At the farthest edges of the solar system lies the Oort Cloud, a hypothetical, distant cloud of icy objects. This mysterious region is believed to be the source of long-period comets, which can take thousands of years to complete a single orbit around the Sun. The Oort Cloud's existence remains a theoretical concept, yet to be observed, representing one of the many mysteries of our solar system awaiting discovery.

3

The Milky Way Galaxy

Understanding Our Galaxy

The Milky Way, a vast and splendid galaxy, is the cosmic home of our solar system. Composed of billions of stars, planets, and other celestial objects, it's a spiral galaxy with a distinct shape and structure. This chapter delves into the composition and characteristics of the Milky Way, helping us understand our place in this immense galaxy.

Structure of the Milky Way

The Milky Way is characterized by its spiral shape, consisting of several arms coiling around a central bulge. These arms, made of dust, gas, and stars, are sites of intense star formation. Our solar system is located in one such spiral arm, known as the Orion Arm, situated about 27,000 light-years from the galactic center. The galaxy's disk, encompassing these spiral arms, is relatively thin compared to its diameter, which spans about 100,000 light-years.

The Galactic Center

At the heart of the Milky Way lies the galactic center, a dense and energetic region. This center is believed to house a supermassive black hole, known as Sagittarius A*, which plays a crucial role in the galaxy's dynamics. The area around the black hole is a hub of activity, with various astronomical phenomena like star clusters, gas clouds, and energetic emissions occurring frequently.

Stars and Nebulae in the Milky Way

The Milky Way is a stellar nursery, hosting an array of stars at different stages of their life cycles. These stars range from tiny red dwarfs to massive blue supergiants. Nebulae, the birthplaces of stars, are abundant in the Milky Way. These clouds of gas and dust come in various forms, such as emission nebulae, which glow brightly, and dark nebulae, which obscure the light of stars behind them. The Orion Nebula and the Horsehead Nebula are two well-known examples.

Life Cycle of Stars

The life cycle of stars is a fascinating aspect of our galaxy. Stars are born in nebulae, where gravity pulls together material to form a new star. Depending on their mass, stars can live for millions to billions of years. Smaller stars, like our Sun, swell into red giants before shedding their outer layers and becoming white dwarfs. Larger stars undergo more dramatic ends, exploding as supernovae, potentially leaving behind neutron stars or black holes. These stellar life cycles play a pivotal role in the distribution of elements throughout the galaxy.

The Milky Way Galaxy is not just a collection of stars and planets; it's a dynamic, living entity, constantly evolving and changing. As we unravel its mysteries, we gain a deeper appreciation for the complexities and wonders of the universe we inhabit.

4

Beyond Our Galaxy

Other Galaxies in the Universe

The universe is a vast expanse filled with countless galaxies, each a magnificent world of its own. Beyond the Milky Way, there are estimated to be over two trillion galaxies, each varying in size, shape, and composition. From spiral galaxies like Andromeda to irregular galaxies like the Magellanic Clouds, the diversity is astounding. This chapter explores the rich tapestry of galaxies that make up our universe.

The Local Group and Neighboring Galaxies

Our Milky Way is part of a galaxy group known as the Local Group, a collection of more than 50 galaxies spread over a diameter of about 10 million light-years. The Andromeda Galaxy, the largest in our group, is on a collision course with the Milky Way, predicted to merge in about 4 billion years. Other notable members include the Triangulum Galaxy and numerous dwarf galaxies. This local neighborhood provides a glimpse into the interactions and dynamics of galaxies in close proximity.

Galaxy Clusters and Superclusters

Galaxies often cluster together under the influence of gravity, forming larger structures known as galaxy clusters. These clusters can contain hundreds to thousands of galaxies. Even larger are superclusters, which are clusters of galaxy clusters. Our own galaxy is part of the Virgo Supercluster, which contains at least 100 galaxy groups and clusters. Superclusters are the largest coherent structures in the cosmos, showcasing the gravitational organization of matter on a massive scale.

Quasars and Black Holes

Quasars are among the most energetic and distant objects in the universe. These bright celestial phenomena are thought to be powered by supermassive black holes at the centers of young galaxies, emitting immense amounts of radiation as matter falls into them. Black holes, fascinating and mysterious, are regions of space where gravity is so strong that nothing, not even light, can escape. They play a crucial role in the formation and evolution of galaxies.

The Expanding Universe

One of the most profound discoveries in cosmology is that the universe is expanding. This expansion, first observed by Edwin Hubble in 1929, implies that galaxies are moving away from each other. The farther away a galaxy is, the faster it appears to be moving away. This expansion is a key aspect of the Big Bang theory, suggesting that the universe started from a singular, dense point and has been expanding ever since.

In this chapter, we have journeyed beyond the confines of our galaxy to explore the grandeur and complexity of the universe. From neighboring galaxies to the vast superclusters, each discovery adds a piece to the puzzle of our cosmic story. As we look further into the universe, we not only explore space but also time, witnessing the history of the cosmos unfold.

5

Fun Facts and Space Trivia

Fun Facts and Space Trivia

Interesting Tidbits About Space

Space is full of surprises and fascinating facts that make it an endlessly intriguing subject. Did you know that a day on Venus is longer than a year on Venus? Or that there are mountains on Pluto made of water ice, which is as hard as rock because of the extremely low temperatures? This section is packed with such interesting tidbits about space, offering a lighter take on the wonders of the cosmos.

Strange Phenomena in the Universe

The universe is home to some truly bizarre and enigmatic phenomena. Imagine a star so dense that just a teaspoon of its material would weigh about a billion tons. Or consider the "Bootes void," an enormous, nearly empty space in the universe, measuring about 330 million light-years across. This section explores these and other strange aspects of the universe that continue to perplex and fascinate scientists and laypeople alike.

Record Breakers in Space

Space is the ultimate frontier for record-breaking feats. From the largest known star, UY Scuti, which is about 1,700 times the size of the sun, to the fastest rotating neutron star, spinning at more than 700 times per second, this section highlights some of the most extreme and astonishing record holders in the cosmos.

Quiz

To engage readers further and test their knowledge, this chapter includes a quiz. Questions range from easy to challenging, covering various aspects of space and astronomy. This fun and educational quiz is a great way for readers to review what they've learned and deepen their understanding of space.

In Chapter 7, we've explored some of the lighter, more intriguing aspects of space and the universe. From quirky facts to mind-boggling phenomena, these tidbits add a sense of wonder and excitement to the study of the cosmos, reminding us of the endless surprises that await us in our exploration of space.

Quiz

Space and Astronomy Quiz
    Easy: What is the closest planet to the Sun?

A) Venus

B) Earth

C) Mercury

D) Mars

Easy: What is the largest planet in our solar system?

A) Jupiter

B) Saturn

C) Neptune

D) Uranus

Moderate: Which planet is known as the "Red Planet"?

A) Mars

B) Jupiter

C) Venus

D) Mercury

Moderate: How many moons does Jupiter have?

A) 4

B) 16

C) 79

D) Over 100

Moderate: What is a supernova?

A) A black hole

B) An exploding star

C) A comet

D) A new star

Challenging: What is the term for the boundary around a black hole beyond which no light or other radiation can escape?

A) Event Horizon

B) Photon Sphere

C) Singularity

D) Accretion Disk

Challenging: Which of the following is the largest type of star in the universe?

A) Red Giant

B) White Dwarf

C) Neutron Star

D) Hypergiant

Challenging: What is the name of the galaxy that contains our Solar System?

A) The Milky Way

B) Andromeda

C) Triangulum

D) Whirlpool

Expert: What is the term for the speed needed to break free from a planet or moon's gravitational pull?

A) Orbital Velocity

B) Escape Velocity

C) Terminal Velocity

D) Light Speed

Expert: Who was the first person to propose that the Sun, not the Earth, was the center of the solar system?

A) Isaac Newton

B) Galileo Galilei

C) Nicolaus Copernicus

D) Johannes Kepler

Answers at end of book.

6

Extraterrestrial Life and Exoplanets

T he Search for Life Beyond Earth

One of the most captivating questions in science is whether life exists beyond Earth. This chapter explores the ongoing search for extraterrestrial life, a quest that has captivated humans for centuries. Advances in technology and space exploration have enabled us to probe deeper into the cosmos, examining planets and moons within our solar system and beyond for signs of life.

Exoplanets: Worlds Beyond Our Solar System

Exoplanets, or extrasolar planets, are planets that orbit stars other than our Sun. The discovery of these distant worlds has revolutionized our understanding of the universe. Thousands of exoplanets have been identified, showcasing an incredible variety of sizes, compositions, and orbits. From rocky Earth-like planets to gas giants larger than Jupiter, the diversity of these worlds challenges our notions of where and how life could exist.

The Goldilocks Zone

The "Goldilocks Zone," also known as the habitable zone, refers to the region around a star where conditions might be just right for life to exist – not too hot and not too cold. Planets in this zone could potentially have liquid water, a key ingredient for life as we know it. The discovery of exoplanets within their star's Goldilocks Zone has sparked excitement about the possibility of finding Earth-like conditions elsewhere in the universe.

Astrobiology: The Science of Alien Life

Astrobiology is the interdisciplinary science that studies the origin, evolution, distribution, and future of life in the universe. This field combines aspects of astronomy, biology, and geology to understand the potential for life on other worlds. Astrobiologists study extreme environments on Earth, such as deep-sea vents and arctic ice, to gain insights into how life might survive under extraterrestrial conditions. The search for microbial life within our solar system, particularly on Mars and the icy moons of Jupiter and Saturn, is a key focus of astrobiological research.

In this chapter, we have explored the fascinating search for life beyond our planet. From the discovery of distant exoplanets to the study of extreme life forms on Earth, each step brings us closer to answering the age-old question: Are we alone in the universe? As we continue to search the stars, the possibility of finding extraterrestrial life remains one of the most thrilling prospects in space exploration.

7

Space Exploration and Technology

History of Space Exploration

This chapter takes you on a journey through the history of space exploration, a tale of human curiosity and technological ingenuity. The space age began in the mid-20th century with the launch of the first artificial satellites and the pioneering flights of astronauts and cosmonauts. This era marked a significant leap in our ability to explore and understand space, transforming science fiction into reality.

Milestones in Space Travel

Space travel has seen numerous milestones, from Yuri Gagarin's first human spaceflight in 1961 to the Apollo moon landings between 1969 and 1972. The Space Shuttle program, spanning from 1981 to 2011, revolutionized space missions with reusable spacecraft. Recently, private companies have joined the space race, opening new avenues for space travel and exploration.

Space Probes and Rovers

Space probes and rovers have been instrumental in exploring distant planets and moons. These unmanned spacecraft have journeyed to the far reaches of our solar system, sending back invaluable data and images. Notable missions include the Voyager probes, which have entered interstellar space, and the Mars rovers, which have provided insights into the Red Planet's geology and potential for past life.

The International Space Station

The International Space Station (ISS) represents a milestone in international cooperation and space research. Orbiting Earth since 1998, the ISS serves as a microgravity laboratory where scientific research is conducted in various fields, including astronomy, biology, and physics. The station has helped us understand the effects of long-duration spaceflight on the human body, vital for future deep-space missions.

Future of Space Exploration

The future of space exploration is as exciting as its past. Upcoming missions aim to return humans to the Moon, explore Mars, and even venture to asteroids and the outer planets. Advances in technology, such as improved propulsion systems and habitats for long-duration missions, are paving the way for the next generation of space exploration. The dream of human settlement on other planets, once a realm of fiction, is slowly becoming a tangible goal.

In this chapter, we have traced the remarkable journey of space exploration, from its early beginnings to its future frontiers. Each step in this journey not only advances our scientific knowledge but also inspires us to dream bigger and reach further into the cosmos.

8

The Mysteries of the Universe

The Big Bang Theory

The Big Bang Theory is the leading explanation about how the universe began. At its simplest, it talks about the universe as we know it starting with a small singularity, then inflating over the next 13.8 billion years to the cosmos that we know today. This chapter delves into the fascinating details of this theory, exploring the initial singularity, the rapid expansion of the universe, and the formation of the fundamental particles and atoms that make up everything we see around us.

Dark Matter and Dark Energy

Despite all our scientific advancements, much of the universe remains mysterious. Dark matter and dark energy are two of the most elusive and intriguing aspects of the cosmos. Dark matter, which does not emit or absorb light, makes up about 27% of the universe. Its presence is inferred from its gravitational effects on visible matter and radiation.

Dark energy, even more mysterious, accounts for about 68% of the universe and is driving its accelerating expansion. This chapter explores what scientists currently understand about these enigmatic components of the universe.

The Puzzle of Time and Space

Time and space are fundamental concepts that have baffled scientists and philosophers for centuries. In the realm of astrophysics, they take on even more complex dimensions. The theory of relativity introduced by Albert Einstein showed that time and space are interwoven into a single continuum known as spacetime. This chapter delves into the intriguing nature of spacetime, black holes, wormholes, and the possibility of time travel, all of which challenge our understanding of the universe.

Theoretical Physics and the Universe

Theoretical physics provides the framework for understanding the universe at its most fundamental level. It encompasses a range of theories, including quantum mechanics, which describes the bizarre world of the very small, and the general theory of relativity, which explains the motion of large objects and the structure of spacetime. This chapter discusses how these theories have shaped our understanding of the universe and the ongoing quest to develop a unified theory that can explain all physical aspects of the universe, including gravity.

In this chapter, we have explored some of the most profound and complex mysteries of the universe. From the origins of the cosmos to the fundamental nature of reality, these mysteries continue to challenge and inspire scientists, driving the quest for deeper understanding and new discoveries.

9

The Moon and Other Natural Satellites

The Moon: Earth's Companion

The Moon, Earth's only natural satellite, has been a source of wonder and inspiration throughout human history. Formed about 4.5 billion years ago, possibly from a collision between Earth and a Mars-sized body, the Moon is a cornerstone in understanding our planet's history. It stabilizes Earth's tilt and climate, influences tides, and has been pivotal in developing our understanding of celestial mechanics. The Moon's surface, marked by craters, mountains, and ancient lava plains, offers clues to the solar system's tumultuous past.

Moons of Other Planets

Beyond Earth, many other moons orbit the planets of our solar system, each with its own unique characteristics. From Jupiter's Galilean moons - Io, Europa, Ganymede, and Callisto - to Saturn's Titan and Enceladus, these moons showcase a stunning diversity. For instance, Io is the most volcanically active body in the solar system, while Europa's icy surface conceals a potentially habitable ocean beneath. Titan, larger than the planet Mercury, boasts a dense atmosphere and liquid methane lakes.

Unique Features of Various Moons

Each moon in our solar system has unique features and mysteries. Mars's moons, Phobos and Deimos, are small and irregularly shaped, likely captured asteroids. Neptune's moon Triton orbits in the opposite direction of the planet's rotation, suggesting it's a captured Kuiper Belt object. Meanwhile, Saturn's moon Mimas resembles the "Death Star" from "Star Wars" due to its large crater, Herschel. These diverse characteristics offer invaluable insights into planetary formation and the evolution of the solar system.

The Role of Moons in Planetary Systems

Moons play significant roles in the dynamics of planetary systems. They can affect planetary rotation, shape climates, and even contribute to the potential habitability of planets. For example, Jupiter's moon Europa is a key target in the search for extraterrestrial life due to its subsurface ocean. The study of moons not only helps us understand the planets they orbit but also gives us a broader understanding of the mechanics and evolution of the solar system.

In Chapter 8, we have explored the fascinating world of the Moon and other natural satellites in our solar system. Each moon, with its unique features and history, adds to the rich tapestry of our cosmic neighborhood. As we continue to explore these diverse and enigmatic celestial bodies, we gain deeper insights into the workings of our solar system and the possibilities that lie beyond.

10

The Cosmic Web and Large-Scale Structure

Understanding the Cosmic Web

The cosmic web is a colossal structure in the universe, consisting of countless galaxies and massive voids, interlinked by filaments of dark matter and gas. This web-like structure is not random; it's a result of the gravitational forces acting over billions of years. Understanding the cosmic web is crucial for comprehending the large-scale structure of the universe and the processes that shaped it.

Filaments and Voids

The cosmic web consists primarily of two components: filaments and voids. Filaments are the largest known coherent structures in the cosmos, massive, thread-like formations made of galaxies and dark matter. They form the boundaries between vast empty spaces known as voids, the dark, nearly empty regions of the universe. These filaments and voids create a pattern similar to a spider's web, with dense clusters of galaxies at the intersections of filaments and vast, empty voids in between.

The Role of Dark Matter

Dark matter plays a pivotal role in the formation and structure of the cosmic web. Although invisible and undetectable by traditional means, dark matter exerts a significant gravitational pull. It's this gravitational force that helped shape the web, pulling galaxies and gas into the filaments and leaving voids in their wake. Understanding dark matter is key to unlocking the secrets of the cosmic web and the evolution of the universe.

Mapping the Large-Scale Structure

Advancements in technology and astronomy have allowed us to map the large-scale structure of the universe in unprecedented detail. Surveys using telescopes and satellites have provided a clearer picture of the distribution of galaxies and dark matter. These maps are crucial for studying the cosmic web and understanding the universe's history, from the Big Bang to the present day.

In Chapter 9, we've explored the fascinating concept of the cosmic web, a fundamental component of the universe's large-scale structure. This web, a network of filaments and voids interwoven with dark matter, holds the key to understanding the universe's past, present, and future. As we continue to map and study this grand structure, we unravel more of the mysteries of the cosmos, taking us one step closer to understanding the vastness of the universe we live in.

11

Astronomical Phenomena

E clipses and Transits

Eclipses and transits are celestial events that have fascinated humans for millennia. An eclipse occurs when one celestial body moves into the shadow of another, like during a solar or lunar eclipse. Solar eclipses happen when the Moon passes between Earth and the Sun, casting a shadow on Earth. Lunar eclipses occur when the Moon moves into Earth's shadow. Transits, on the other hand, involve a celestial body passing directly between a larger body and the observer – like the Transit of Venus, where Venus crosses in front of the Sun as seen from Earth. These events are not just spectacular to witness but also offer valuable opportunities for scientific research.

Supernovae and Neutron Stars

Supernovae are among the most energetic events in the universe. They occur at the end of a star's life cycle, resulting in a massive explosion that can outshine entire galaxies. Supernovae play a key role in distributing elements throughout the universe. The remnants of these explosions can result in the formation of neutron stars – incredibly dense cores left behind after a supernova. Neutron stars, with masses greater than the Sun but only a few kilometers in diameter, are some of the densest objects in the universe.

Pulsars and Magnetars

Pulsars are a type of neutron star that emit beams of radiation from their magnetic poles. As pulsars rotate, these beams sweep across the sky, and if aligned with Earth, can be observed as regular pulses of radiation. Magnetars are a rare type of neutron star with extremely powerful magnetic fields, millions of times stronger than any made by humans. These intense fields can cause starquakes and powerful bursts of X-rays and gamma rays.

Gravitational Waves

Gravitational waves are ripples in the fabric of spacetime caused by some of the most violent and energetic processes in the universe, such as colliding black holes or exploding stars. Predicted by Einstein's theory of general relativity, gravitational waves were first directly detected in 2015. This discovery opened a new window in astronomy, allowing scientists to observe cosmic events that were previously undetectable and providing new insights into the nature of gravity and the universe.

In Chapter 10, we've delved into some of the most awe-inspiring and complex astronomical phenomena. From the breathtaking beauty of eclipses and transits to the powerful forces of supernovae and gravitational waves, these phenomena not only capture our imagination but also offer invaluable insights into the workings of the universe. As we continue to observe and study these events, we deepen our understanding of the cosmos and our place within it.

12

Observing the Universe

Telescopes and Observatories

Telescopes are the primary tools for observing the universe, acting as windows to the cosmos. This chapter begins by exploring the evolution of telescopes, from Galileo's first refracting telescope to the sophisticated observatories of today. Ground-based observatories, like the Very Large Telescope in Chile and the Mauna Kea Observatories in Hawaii, are equipped with advanced technologies that allow astronomers to observe distant galaxies, exoplanets, and other celestial phenomena. The functioning, location, and advancements of these observatories highlight how they have become instrumental in our understanding of the universe.

Space Telescopes and Their Discoveries

Space telescopes, free from the distortions of Earth's atmosphere, have revolutionized our view of the universe. The Hubble Space Telescope, for instance, has provided some of the most detailed images of distant galaxies, nebulae, and star clusters. The Kepler Space Telescope significantly advanced the search for exoplanets. This section delves into the missions, discoveries, and contributions of various space telescopes, emphasizing their role in expanding our cosmic knowledge.

Amateur Astronomy

Amateur astronomy is a popular and valuable aspect of astronomical observations. Equipped with consumer-grade telescopes and other equipment, amateur astronomers have contributed significantly to the field, including discovering comets, monitoring variable stars, and capturing transient events. This section explores how amateur astronomy complements professional research and offers guidance on how enthusiasts can get involved in this exciting and rewarding hobby.

The Evolution of Astronomical Observations

The final part of this chapter looks at the history and future of astronomical observations. From early stargazing with the naked eye to the development of telescopes and the advent of digital imaging and spectroscopy, astronomical observations have come a long way. We explore how these advancements have deepened our understanding of the universe and look forward to future technologies, like the James Webb Space Telescope and ground-based Extremely Large Telescopes, which promise to further unveil the mysteries of the cosmos.

In Chapter 11, we've journeyed through the history and future of observing the universe. From the earliest telescopes to the most advanced observatories and space telescopes, each step in this evolution has brought us closer to understanding the vastness and complexity of the cosmos. As we continue to improve our observational capabilities, the universe reveals more of its secrets, beckoning us to keep watching and exploring.

13

More Facts and Answers

F acts

1. Black holes can spin at nearly the speed of light.
2. The Milky Way galaxy will collide with the Andromeda Galaxy in about 4 billion years.
3. The largest known star, UY Scuti, could fit more than 1,700 Suns in it.
4. Neptune was the first planet to get its existence predicted by mathematical calculations before it was actually seen through a telescope.
5. The sun makes up around 99.86% of the solar system's mass.
6. The ice deposits found at Mars' poles contain as much water as Lake Superior, Earth's largest freshwater lake.
7. Venus is the only planet to spin clockwise.
8. A year on Venus is shorter than its day - it takes longer to rotate on its axis than it does to orbit the sun.
9. The dwarf planet Haumea has an elongated shape due to its rapid rotation.

10. Saturn's moon Enceladus has geysers that shoot water vapor and ice particles into space.

11. Mercury's orbital period around the sun is 88 days, making its year shorter than its day.

12. Pluto's orbit sometimes brings it closer to the Sun than Neptune.

13. Jupiter's moon Io has over 400 active volcanoes, making it the most geologically active object in the solar system.

14. The Crab Nebula is the result of a supernova explosion that was observed on Earth in 1054.

15. The Horsehead Nebula is part of the larger Orion Nebula, one of the most photographed objects in space.

16. The Oort Cloud is a theoretical cloud of predominantly icy planetesimals proposed to surround the Sun at distances ranging from 2,000 to 200,000 AU.

17. The asteroid belt contains billions of asteroids.

18. The largest moon in the solar system, Ganymede, is larger than the planet Mercury.

19. The largest mountain in our solar system, Olympus Mons on Mars, is three times the height of Mount Everest.

20. The solar system is 4.6 billion years old.

21. The Voyager 1 spacecraft, launched in 1977, is the farthest human-made object from Earth.

22. The temperature on the surface of the Sun is about 5,500 degrees Celsius.

23. The Great Red Spot on Jupiter is a giant storm that has been raging for at least 350 years.

24. The Sun contains more than 99.8% of the total mass of the Solar System.

25. Mercury is the smallest planet in the Solar System.

26. Earth's atmosphere extends to a distance of 10,000 km.

27. Saturn has the most extensive ring system of any planet in the Solar System.

28. The first spacecraft to visit Mars was Mariner 4 in 1965.

29. The Kuiper Belt is a region of the Solar System beyond the planets, extending from Neptune's orbit.

30. The Valles Marineris canyon on Mars is 10 times longer and 5 times deeper than the Grand Canyon.

31. The largest diamond known in the universe is a star named Lucy, a crystallized white dwarf.

32. The Sun travels around the galaxy once every 225-250 million years - a journey known as the galactic year.

33. The largest known structure in the Universe is the Hercules-Corona Borealis Great Wall, an enormous galactic supercluster.

34. The distance between stars in the sky is so vast that traveling at the speed of light would still take thousands of years to reach them.

35. The Helix Nebula is sometimes referred to as the "Eye of God" due to its eye-like appearance.

36. The Milky Way is part of a cluster of galaxies known as the Local Group.

37. The Cat's Eye Nebula is one of the most complex nebulae known.

38. The Perseid meteor shower, one of the brightest meteor showers, occurs every year in August.

39. The Earth's magnetic field helps protect us from the harmful effects of solar radiation.

40. Uranus was originally called George's Star after King George III.

41. The Hubble Space Telescope can observe stars and galaxies up to 13.4 billion light years away.

42. The first artificial satellite, Sputnik, was launched by the Soviet Union in 1957.

43. A light-year is about 9.5 trillion kilometers (5.9 trillion miles).

44. The first living mammal to go into space was a dog named Laika from Russia.

45. The sun is classified as a G-type main-sequence star, also known as a yellow dwarf.

46. There are estimated to be over 100 billion galaxies in the universe.

47. The Methuselah star, HD 140283, is nearly as old as the universe itself, at 14.5 billion years.

48. The Pillars of Creation, part of the Eagle Nebula, are large towers of cosmic dust and gas where new stars are being born.

49. The Drake Equation is used to estimate the number of active, communicative extraterrestrial civilizations in the Milky Way galaxy.

50. The Leonids meteor shower results from the Earth passing through the debris left by the comet Tempel-Tuttle.

51. The term "astronaut" comes from Greek words meaning "star" and "sailor."

52. The first confirmed detection of exoplanets was made in 1992.

53. The Sloan Great Wall is one of the largest known structures in the universe, a giant wall of galaxies measuring 1.38 billion light-years in length.

54. The Andromeda Galaxy is the largest galaxy in the Local Group, which also includes the Milky Way, the Triangulum Galaxy, and about 54 other smaller galaxies.

55. The Boomerang Nebula is the coldest known place in

the universe, with temperatures even lower than the background radiation of space.

56. The age of the universe is approximately 13.8 billion years.
57. The International Space Station orbits the Earth every 90 minutes.
58. The sun's energy output every second is equivalent to about 4 million times the world's energy consumption.
59. The oldest known star in the universe is SMSS J031300.36-670839.3, believed to be around 13.6 billion years old.
60. Earth is the only planet not named after a god.
61. The moon is the fifth largest natural satellite in the Solar System.
62. The universe is thought to be flat.
63. The first space shuttle was Columbia, launched on April 12, 1981.
64. The Laniakea Supercluster is the galaxy supercluster that is home to the Milky Way.
65. The largest known black hole, TON 618, is estimated to have a mass of 66 billion times that of the Sun.
66. The speed of light in a vacuum is about 299,792 kilometers per second (186,282 miles per second).
67. The Apollo 11 mission was the first manned mission to land on the Moon.
68. The temperature in the core of a supernova explosion can reach 100 billion degrees Celsius.
69. The cosmic microwave background radiation is the after-glow of the Big Bang.
70. The Kepler Space Telescope was launched to find Earth-size planets orbiting other stars.
71. The Van Allen radiation belts are two layers of charged particles that surround Earth, held in place by Earth's

magnetic field.

72. The Alpha Centauri star system is the closest to the Solar System, at 4.37 light-years away.

73. The Tarantula Nebula is the most active star-forming region in our local group of galaxies.

74. The Rosetta spacecraft was the first to orbit and land on a comet.

75. The Andromeda Galaxy and the Milky Way are expected to collide in about 4 billion years.

76. The largest moon in the Solar System, Ganymede, is larger than the planet Mercury.

77. The dwarf planet Eris was one of the reasons Pluto was demoted from being a planet.

78. The Great Attractor is a gravitational anomaly in intergalactic space that is drawing galaxies towards it.

79. The age of the Solar System is about 4.6 billion years, the same age as the Earth.

80. The Orion Nebula is one of the brightest nebulae, visible to the naked eye in the night sky.

81. The Large Hadron Collider is the world's largest and most powerful particle accelerator.

82. The Tunguska event in 1908 was caused by an asteroid or comet fragment exploding in Earth's atmosphere.

83. The Wow! signal was a strong radio signal received by astronomers in 1977 that many speculate could have been from an extraterrestrial source.

84. The Higgs boson, sometimes referred to as the "God particle," was discovered at the Large Hadron Collider in 2012.

85. The most distant known object in the Solar System is the Farout, which is more than 100 times farther from the Sun

than Earth is.

86. The Chelyabinsk meteor in 2013 was the largest known natural object to have entered Earth's atmosphere since the Tunguska event.

87. The Sun accounts for about 99.86% of the Solar System's total mass.

88. The Phoenix Mars Lander discovered water ice on Mars in 2008.

89. The first image of a black hole was captured by the Event Horizon Telescope in 2019.

90. The universe is thought to be composed of 68% dark energy, 27% dark matter, and 5% ordinary matter.

91. The largest impact crater in the Solar System is the Hellas impact basin on Mars.

92. 92.The dwarf planet Ceres, located in the asteroid belt, contains water ice.

93. The Andromeda Galaxy and the Milky Way are on a collision course in about 4 billion years.

94. There are at least five dwarf planets recognized in our solar system.

95. The largest dwarf planet in the solar system is Eris.

96. Neptune was the first planet to be found through mathematical prediction rather than observation.

97. The Voyager 1 spacecraft is the most distant human-made object from Earth.

98. Neptune's winds are the fastest in the solar system, reaching speeds of 2,100 km/h.

99. The highest mountain known in the universe is Olympus Mons, located on Mars.

100. The first known asteroid was Ceres, discovered in 1801.

101. Saturn's largest moon, Titan, has a very thick atmosphere

- thicker than Earth's.

102. The Oort Cloud is thought to surround our solar system at a distance of up to one light-year.

103. The Crab Nebula was created by a supernova explosion witnessed in 1054.

104. The Horsehead Nebula is one of the most identifiable nebulae because of its unique shape.

105. The largest structure found in the universe is the Hercules-Corona Borealis Great Wall, a galactic filament about 10 billion light-years across.

106. The oldest known star is 13.2 billion years old.

107. The Perseids and the Geminids are two of the most spectacular annual meteor showers.

108. The space between galaxies is not empty but filled with a sparse intergalactic medium.

109. The largest known diamond in the universe is a crystallized white dwarf star named Lucy.

110. The Great Attractor, a gravitational anomaly in intergalactic space, is drawing galaxies towards it.

111. The Boomerang Nebula is the coldest place in the universe, with temperatures even lower than the background radiation.

112. The universe's largest known void is the Giant Void, which is nearly 1.3 billion light-years across.

113. The Methuselah star, a star in our galaxy, appears to be older than the universe.

114. The sun's magnetic field is so strong that it can affect the entire solar system.

115. The term "space" was first used to describe the region beyond Earth's sky in the 14th century.

116. "Shooting stars" are actually not stars but meteors burning

up upon entering Earth's atmosphere.

117. The first photograph of Earth from space was taken in 1946.

118. The first animal to go into space was a dog named Laika from Russia.

119. The Cat's Eye Nebula was one of the first planetary nebulae to be discovered.

120. A year on Uranus lasts about 84 Earth years.

121. Jupiter's moon Io is the most volcanically active body in the solar system.

122. The first successful Mars rover was Sojourner, part of the Mars Pathfinder mission in 1997.

123. The first confirmed Earth-sized exoplanet in a star's habitable zone was Kepler-186f.

124. Astronauts cannot burp in space – there is no gravity to separate liquid from gas in their stomachs.

125. The first satellite, Sputnik, was launched by the Soviet Union in 1957.

126. The largest solar flare ever recorded happened on November 4, 2003.

127. The first American to orbit the Earth was John Glenn in 1962.

128. The first food eaten in space by an American astronaut was applesauce.

129. The cosmic microwave background is radiation left over from the Big Bang.

130. If you could compress Earth down to the size of a marble, it would collapse on itself and become a black hole.

131. There are more solar systems in the universe than there are grains of sand on all the beaches on Earth.

132. Black holes can spin at nearly the speed of light.

133. The dwarf planet Haumea has an orbit shaped like an elongated ellipse.
134. The Pillars of Creation are so named because the gas and dust are in the process of creating new stars.
135. The Voyager probes are the farthest human-made objects from Earth and carry a golden record with sounds and images of Earth.
136. The first man to walk on the moon was Neil Armstrong in 1969.
137. The first woman to command a space shuttle mission was Eileen Collins.
138. The Hubble Space Telescope was launched into orbit in 1990.
139. The first multi-person crew to orbit the Earth was on the Voskhod 1 mission in 1964.
140. The largest solar system object beyond Neptune is Eris.
141. The sun's solar flare can extend up to several times the size of Earth.
142. There are likely billions of other solar systems in our galaxy.
143. The first space station was the Soviet Salyut 1, launched in 1971.

Chapter 5 Answers

1)Easy: What is the closest planet to the Sun?

Answer: C) Mercury

2)Easy: What is the largest planet in our solar system?

Answer: A) Jupiter

3)Moderate: Which planet is known as the "Red Planet"?

Answer: A) Mars

4)Moderate: How many moons does Jupiter have?

Answer: C) 79 (As of my last update in April 2023; this number could have changed if more moons were discovered since then.)

5)Moderate: What is a supernova?

Answer: B) An exploding star

6)Challenging: What is the term for the boundary around a black hole beyond which no light or other radiation can escape?

Answer: A) Event Horizon

7)Challenging: Which of the following is the largest type of star in the universe?

Answer: D) Hypergiant

8)Challenging: What is the name of the galaxy that contains our Solar System?

Answer: A) The Milky Way

9)Expert: What is the term for the speed needed to break free from a planet or moon's gravitational pull?

Answer: B) Escape Velocity

10)Expert: Who was the first person to propose that the Sun, not the Earth, was the center of the solar system?

Answer: C) Nicolaus Copernicus

14

Conclusion

As we conclude our cosmic journey with "Space Facts Galore," it's essential to reflect on the incredible vastness and complexity of the universe we inhabit. This book has taken us from the familiar landscapes of our solar system to the enigmatic depths of distant galaxies, unveiling a universe that is both awe-inspiring and humbling.

Reflecting on the Vastness of the Universe

The universe, with its billions of galaxies and trillions of stars, is a testament to the grandeur and mystery of the cosmos. Each chapter of this book has revealed just how expansive and varied the universe is. From the birth of stars in distant nebulae to the intricate dance of galaxies, we've glimpsed the universe's vastness and its many wonders. This reflection not only deepens our appreciation for the cosmos but also our understanding of our place within it. We are but a small part of an immense and ever-expanding universe, a thought that is both sobering and exhilarating.

The Future of Space Research and Discovery

Looking ahead, the future of space research and discovery holds limitless possibilities. As technology advances, so does our ability to explore deeper into the cosmos. Upcoming missions to Mars, the study of exoplanets, and the ongoing search for extraterrestrial life are just the beginning. The pursuit of knowledge about our universe continues to drive innovation, leading to advancements that benefit not only space exploration but also life on Earth. The journey into space is far from over; it is continually evolving, with each new discovery propelling us further into the unknown.

Inspiring the Next Generation of Explorers

Perhaps one of the most significant impacts of space exploration is its ability to inspire. This book aims to ignite a spark of curiosity and wonder in its readers, especially the young minds who will become the next generation of astronomers, astronauts, and astrophysicists. The stories of space exploration and the mysteries of the cosmos are not just scientific endeavors; they are also adventures that call to the explorer in each of us. It is our hope that "Space Facts Galore" will encourage readers to look up at the night sky with a sense of wonder and a desire to uncover the universe's secrets.

In closing, "Space Facts Galore" has been a journey through the incredible expanse of space, a reminder of the universe's beauty, complexity, and endless possibilities. As we stand on the brink of new discoveries and explorations, the universe continues to be a source of inspiration and wonder, challenging us to keep looking beyond the horizon and dreaming about what lies in the far reaches of the cosmos.